An Overview of Free Energy

Vinyasi – 20 July, 2021

Mho's Law is not a law per se, but is a natural derivation of Ohm's Law if – per chance, for about a moment – we forsake our allegiance and blind obedience (of faith) towards the Conservation of Energy pseudo-law and everything else of similar design and intention (acting as baggage) that could bring us down into a very narrow view of reality.

This is due to the fact that the Conservation of Energy wannabe-a-law is not a mathematical relationship. Strictly speaking, it is a colloquialism in which we paraphrase, very loosely, the expression in which "energy entering into a circuit has to equal the energy which exits out of a circuit." But nowhere, within the field of electrodynamic theory, will we find a mathematical equation that specifically states this relationship. Instead, we'll find that Ohm's Law specifically equates the power which exits a circuit must equal the voltage squared applied to the circuit (coming from outside the circuit and acting as the power supply for the circuit) divided by the resistance within the circuit. This is where the Conservation of Energy comes from. But it only pertains to the consumption, namely the conversion of electrical energy within our appliances, into formats of energy other than electrical energy, such as: heat or light or mechanical motion of an electric motor. It has nothing to do with the production of energy. It only pertains to its consumption or conversion which amounts to the same thing.

It may be said of Ohm's Law, and of the Conservation of Energy pseudo-law, that this relationship of energy, which enters a circuit and energy which exits a circuit, is a symmetrical relationship due to the equivalence of energy's entrance and exit must exactly equal each other, including any minor losses due to leakage and waste.

Some producers of energy comply with this Conservation of Energy perspective in as much as they are non-reactive generators of energy, such as: batteries. But not all producers of energy are governed exclusively by Ohm's Law, because Ohm's Law only pertains to real power. It does not govern the behavior of reactive power. This is where the derivative of Ohm's Law, suggested to us by Lord Kelvin more than a century ago,[1] comes into play.

But wouldn't you know it, the international body of people who are in charge of the units of measurement have swept aside Lord Kelvin's suggestion of our use of the word Mho, which is: Ohm spelled backwards, as the symbol which represents the unit of measurement for conductivity - also known as admittance,[2] in favor of the use of the word: Siemens to replace Mho. And to further their endeavor to make free energy and overunity go away and disappear (by causing us to misunderstand energy, much less understand free energy), they have associated the temperature of absolute zero as a unit of measurement to be named after Lord Kelvin, himself, as if to suggest (by way of implication) that the absolute freezing temperature of zero degrees Kelvin is the only way to create super-conductivity and, thus, create overunity and free energy when (in fact) this is not entirely true!

It is true, in a limited sense (in the sense that we could reduce the temperature of a piece of wire, or a coil of wire, to reduce its resistance to the flow of current by a quantity which approaches zero Ohms of resistance). But why do things the hard way when we can do things the easy way and listen to Lord Kelvin's suggestion and take him seriously!?

Before I go any further, I should sweep aside another convention (of collective naiveté) which is another fiction of our own creation intended for a good reason: to make the life and work of the electrical engineer a little easier by mathematically simplifying Ohm's Law by substituting the capital letter "I" in place of voltage divided by resistance and multiply that back into the remaining voltage that will result in the wattage of power that exits a circuit. But this is a short hand – a mathematical contrivance – that is not a reality. In reality, all we have (under Ohm's Law) is a change in voltage squared within time brought about by its division by the resistance within a circuit.

Thus,...

The length of a piece of wire defines the voltage (dielectric potential) of that wire.

---

[1] Siemens (unit), Mho → https://is.gd/tidose = https://en.wikipedia.org/wiki/Siemens_(unit)#Mho

[2] Admittance → https://is.gd/fupene = https://en.wikipedia.org/wiki/Admittance

Thus, is Ohm's Law defined by the geometry of a piece of wire!

$$\text{Power} = \frac{\text{Voltage}^2}{\text{Resistance}}$$

one-dimensional voltage across the length of a piece of wire = V

Voltage squared defines a cross-sectional slice through a piece of wire at right-angles to its length. This implies resistance is a two-dimensional phenomenon based on the diameter of the wire. The length of a piece of wire possesses no diameter. It merely possesses length. Thus, its length possesses no resistance. The length of a piece of wire merely possesses a time-delay which compounds the result of resistance applied to a length of wire extended over its consequence of a change in voltage per unit of time. Thus, is the fiction of current born of this time-delay in which it takes time for information to travel down the length of a piece of wire & for the intelligence of the copper wire to respond with an appropriate action showing it possesses life!

The cross-sectional slice through the width of that wire, and perpendicular to its length, defines its current by virtue of the resistance which the square area of this cross-section defines. This square area brings about a squaring of its voltage divided by (impeded by; resisted by) its resistance.

Current exists as an artifact of time-delay resulting from the rate-of-change of voltage applied over the length of a piece of wire. As such, current is a mathematical fiction born of an over-simplification.

In reality, there is no parameter, called: "current" within Ohm's Law. All that exists is...

$$Power = \frac{Voltage^2}{Resistance} \qquad §1.1a$$

What makes this relationship, of §1.1a, practical is the relation of §1.1b...

$$Power = Voltage \times Current \qquad §1.1b$$

...which has resulted from the extraction of §1.2a...

$$Current = \frac{Voltage}{Resistance} \qquad §1.2a$$

...from equation §1.1a, above, by way of its substitution (and simplification)...

$$Power = \frac{Voltage^2}{Resistance} = Voltage \times \frac{Voltage}{Resistance} = Voltage \times Current \qquad §1.1ab$$

...to create §1.1b as a shorthand version of §1.1a which makes the following relationships possible as a consequence, besides §1.2a...

$$Resistance = \frac{Voltage}{Current} \qquad §1.2b, \text{ and...}$$

$$Voltage = Resistance \times Current \qquad §1.2c$$

Anything other than watts, namely: other than real power, is purely informational in the form of a measurement of volts versus a measurement of amperes, called volts/amperes (VA), which is not considered to be energy, per se, but is considered to be reactive power: a fragmentation of power into its constituent ingredients of magnetism and dielectric potential per units of duration.

The resistance of Ohm's Law defines power (measured in watts) whenever the polarity of current matches the polarity of voltage (ie, positive polarity of voltage versus positive polarity of current; or, negative polarity of voltage versus negative polarity of current). Current flows from areas of high voltage towards areas of lower voltage under Ohm's Law.

The relation which is inversely related to (the mathematical reciprocal, or multiplicative inverse, of) power is the relationship of admittance (conductivity, G, measured in Siemens; formerly measured in units of "mho", ℧). This had been named: Mho by Lord Kelvin, before it was superseded by Siemens, and I endorse and will revive the use of Mho as a Law for the purposes of this discussion, in which the polarity of voltage is opposed to the polarity of current. So, whenever the voltage of conductivity has a polarity of positive sign value, then the polarity of the current of conductivity is signed negative. And whenever the voltage of conductivity has a polarity of a negative sign value, then the polarity of the current of conductivity is signed positive. This effectively inverts voltage so that current flows from areas of low voltage towards areas of higher voltage creating a condition which has colloquially come to be known as: negative resistance[3] (although, as we'll see in a minute, it is more accurately {puritanically} described as being negative voltage)...

$$Conductivity = \frac{Resistance}{-Voltage^2} \qquad §2.1a$$

What makes this relationship, of §2.1a, practical is the relation of §2.1b...

$$Conductivity = \frac{Current}{Voltage\sqrt{-1}} \qquad §2.1b$$

...which has resulted from the extraction of §2.2a...

$$Current = \frac{Resistance}{Voltage\sqrt{-1}} \qquad §2.2a$$

...from equation §2.1a, above, by way of its substitution (and simplification)...

---

3 Negative resistance → https://is.gd/negresist = https://en.wikipedia.org/wiki/Negative_resistance

$$Conductivity = \frac{Resistance}{-Voltage^2} = \frac{1}{Voltage\sqrt{-1}} \times \frac{Resistance}{Voltage\sqrt{-1}} = \frac{Current}{Voltage\sqrt{-1}} \qquad §2.1ab$$

...to create §2.1b as a shorthand version of §2.1a which makes the following relationships possible as a consequence, besides §2.2a...

$$Resistance = Voltage\sqrt{-1} \times Current \qquad §2.2b, and...$$

$$Voltage\sqrt{-1} = \frac{Resistance}{Current} \qquad §2.2c$$

Ohm's Law defines the symmetry of entropic thermodynamics, namely: the symmetry of the Conservation of Energy (which should be renamed: the Conservation of Consumption), in which the volts and the amperes are both real numbers and their multiplication with each other will always result in a positively signed outcome implying the Consumption of Energy.

Electronic simulators don't have Mho's Law built into their design. Their engineers have assumed that Ohm's Law defines everything due to their assumption that entropy defines everything and under all circumstances, including the generation of reactive power. Thus, do they impose the presumption that the Conservation of Energy applies to all circumstances and Mho's Law does not exist as a viable option to compliment Ohm's Law. These presumptions are due to engineers assuming that external prime movers (ie, prime movers which are outside of circuits contributing their energy as an input towards the circuit's outcome) are always needed to engage the generation of reactive power and that circuits cannot, or should not be allowed to, do this on their own (acting as their own prime mover) and should not be allowed to demote the use of externalized voltage inputs to the status of mere stimulants. Stimulants merely motivate this process (acting as a catalyst) and encourage the circuit to avoid its exclusive dependency upon external support due to the fact that stimulants are defined by Mho's Law as having the greatest impact whenever a circuit's input voltage is severely reduced while taking advantage of an increased resistance and an increased impedance (whenever the inversion {negation} of voltage occurs) which actually favors a beneficial outcome of increasing conductivity and the overunity of a circuit's output through the use of the following relationship inherent within Mho's Law and derived from Ohm's Law...

$$Conductivity = \frac{Resistance}{-Voltage^2} \qquad \S 2.1a$$

Mho's Law defines the asymmetry of negentropic thermodynamics, namely: the asymmetry of the Production of Energy, in which the volts and the amperes are both imaginary numbers and the division of resistance by an imaginary voltage yields a current which, when divided by an imaginary voltage (again, to create a squared voltage) yields a negative conductivity and implies the Production of Energy via enhanced conductivity at room temperature (without the need to supercool anything to nearly absolute zero degrees Kelvin)...

$$Conductivity = \frac{Resistance}{-Voltage^2} = \frac{1}{Voltage\sqrt{-1}} \times \frac{Resistance}{Voltage\sqrt{-1}} = \frac{Current}{Voltage\sqrt{-1}} \qquad \S 2.1ab$$

Poor Lord Kelvin is probably squirming in his grave due to nobody is seriously taking his suggestion of utilizing the conductivity, and the super-conductivity at room temperature which Mho's Law is capable of, as the complimentary concept to the resistivity of Ohm's Law. Instead, his suggestion has been replaced by naming absolute zero degrees temperature after him rather than taking his advice and actively pursue super-conductivity the easy way! Instead, we pursue super-conductivity the hard way making it difficult for the common man to benefit from cheap and readily available energy.

Mho's Law lays the foundation for free energy's existence. This is the reason why it has fallen out of favor for use by engineers and scientists, because it justifies free energy and this is against the dictates of industry having a monopoly on energy.

This type of industrial cartel was mentioned and described at length by President Eisenhower during his farewell address to the nation when his term of office was about to expire on the 17th of January 1961.[4] This cartel consists of an extremely binding relationship between commerce and military to do whatever it takes to further their mutual goals of the monopolization of energy and information. This requires governmental control, and

---

4 Eisenhower's Farewell Address to the Nation, as a PDF text → https://is.gd/ahuleq = https://americanrhetoric.com/speeches/PDFFiles/Dwight%20D.%20Eisenhower%20-%20Farewell%20Address.pdf or as an MP3 audio → https://is.gd/opohug = https://americanrhetoric.com/mp3clips/politicalspeeches/dwighteisenhowerfarewell.mp3

commercial control, over energy and information and entertainment to exclude whatever truths could jeopardize their cartel. They have effectively disenfranchised us from transcending our exclusive dependency upon their authority.

Imagine how formidable a task it is to increase energy if voltage is a limited resource (which it is under Ohm's Law) and resistance is always excessively getting in the way of getting any power out of a circuit. Yet, if Mho's Law authorizes the excessive production of power, *not despite* the presence of resistance and the lack of voltage, but *requiring* these two conditions which – would be limiting conditions under Ohm's Law, yet – are encouraged under Mho's Law.

This is why Ohm's Law doesn't work very well for the generation of power, yet, describes the consumption of power very aptly. It is Mho's Law which describes the efficient generation of power at an extremely low cost to its operator. And it is the reactance formula,[5] of capacitance, inductance and frequency (and the use of: 2π), which describes the regulation of free energy spawned and authorized by Mho's Law.

We have had the dexterity of Mho's Law swept under the proverbial rug of ignorance by the substitution of Ohm's Law by its replacement with the Conservation of Energy Law. It is high time we forget about the craftiness of the Conservation of Energy Law in its ability to oversimplify the situation. Let us revive Mho's Law in partnership with Ohm's Law for a complete perspective on energy.

Current is a term designating a mathematical shorthand operating upon voltage versus resistance...

$$Current = \frac{Voltage}{Resistance} \qquad \S1.2a$$

...and...

$$Current = \frac{Resistance}{Voltage\sqrt{-1}} \qquad \S2.2a$$

---

5 Electrical Reactance – Wikipedia → https://is.gd/olavaf = https://en.wikipedia.org/wiki/Electrical_reactance

And because the voltage of Mho's Law is both the multiplicative inverse of voltage (originally derived under Ohm's Law) as well as its additive inversion of signed polarity (positive voltage inverted into negative voltage, or else negative voltage inverted into positive voltage), then (consequently) the current which arises as a form of shorthand notation for these two laws takes on two qualities of: conventional (real) current (under Ohm's Law) possessing the same polarity of sign as does voltage yielding a positive wattage, while reactive current (under Mho's Law) possesses a polarity of sign which is inverse to voltage. It is this latter condition of current, reactive current, which defines the generation of reactive power emanating from out of a voltage source, such as: a battery, or a rotary generator. Conventional current, on the other hand, is restricted to defining the consumption of real power and adheres to the Conservation of Energy dictum, namely: that the energy which enters into an electronic component (which is engaging in the consumption and conversion of this electrical energy) must equal the energy which results from this conversion, such as: the heat arising from a resistor, or the mechanical motion of an electric motor, etc, consequently: "energy IN must equal energy OUT."

Now do you understand the limited jurisdiction of the Conservation of Energy Law?

It's limited to the energy which enters into any electronic component which is engaging in the consumption of energy, namely: its conversion into some other format, such as: the conversion of electrical energy into heat or rotary motion, versus the heat or rotary motion which exits that component. So, if a resistive element has X units of electricity entering that resistor, then X units of heat must exit that resistor – no more and no less. That's it....that's as far as the limitations of physics can take their precious law of Conservation to and not proceed any further with it.

Equation §2.2a has been extracted (subtracted) from equation §2.1a to yield equation §2.1b....

$$Conductivity = \frac{Resistance}{-Voltage^2} = \frac{1}{Voltage\sqrt{-1}} \times \frac{Resistance}{Voltage\sqrt{-1}} = \frac{Current}{Voltage\sqrt{-1}} \qquad \S2.1ab$$

If we hadn't done this and kept conductivity equaling the square of voltage having an inverse (negative) relationship with resistance and avoid the convenience of artificially creating the mathematical construct (ie, abstraction; pseudo-fiction) of current, then we wouldn't be dealing with an imaginary value for the super-conductive variety of current under Mho's Law (in contrast to the resistive variety under Ohm's Law). Instead, we'd be dealing (merely) with a negative square of voltage and be able to create a net total of summing the two subtotals of: reactive power (conductivity; generation), plus real power (resistivity; consumption) to arrive at an awareness as to whether or not a circuit's segregated analysis is symmetrically thermodynamic, overall, obeying the Conservation of Energy by automatically "balancing the load" in which reactive power generation equals the consumption of real power, or else it is deviating from this convention by being asymmetrically non-thermodynamic (negentropic or entropic) in which its generation of reactive power is greater than, or less than, its consumption of real power.

Electronic simulators won't tell us whether or not an electronic component is producing reactive power (acting as a generator). Yet, they still tell us whether or not a component's voltage is positive or negative and whether or not this same component's current is positive or negative and, thus, will tell us whether or not their product is positive or negative. Thus, it doesn't matter that we've confused the situation by ignoring Mho's Law since we can steer clear of this confusion with a proper understanding of what is *really* happening by reeducating ourselves on the significance of Mho's Law and what this has to offer in the way of explaining, and justifying, free energy.

What is *really* happening is that current travels towards areas of lesser voltages only within the domain of electronic components which are acting as consumers of real power, because it is only these components which are symmetrically obeying the Conservation of Energy. Meanwhile, current travels towards areas of greater voltages only within the domain of electronic components which are acting as producers of reactive power, because it is only these components which are asymmetrically obeying Mho's Law and the generation of reactive power which lies outside the jurisdiction of the Conservation of Energy. *{It is this latter condition which accentuates voltage differences, rather than equalizing them, which makes* *possible the accumulation of reactive power achieving infinite levels of amplitude.}*

In other words, current does not travel between and among components of a circuit. Being a mathematical construct, the domain of the traversal of current is strictly within the domain of the component to which this traversal is attached. Only voltage differences exist between components of a circuit. And only resistances exist within components of a circuit. Nothing else matters whenever seeking a tabulation of power and a segregated mapping intended to analyze what is happening.

Because of this pseudo-fictional creation of a mathematical construct, current has made it possible for us to take this construct one step further and misunderstand the situation so completely that we no longer understand energy much less understand free energy – in other words, we fail to understand and appreciate the limitations of real power versus freely available, reactive power.

Current is an artifice, an artificial construct, spawned by the mind of the mathematician intended to simplify the squaring of voltage for both Ohm's Law and Mho's Law. This is analogous, although not equivalent, to the mathematical pseudo-fiction of complex numbers. These mathematical constructs help to simplify the perspective of the electrical engineer performing the calculations of electrical engineering.

Since current does not exist in any ultimate sense, neither does the motion of the electron exist except as a mathematical resultant of changes made to the levels of voltages at various locations in space. In other words, current is a derivative – not a fundamental property – of voltage and resistance. "Movement" is a fiction while "change" is not. The *apparent* existence of movement is what our brain wants to believe is true without any "a priori" foundation to its existence, but with an "a posteriori" authenticity derived from a lack of clarity and honesty. A study of these mathematics – of Ohm's Law and Mho's Law – reminds us of the Vedic perspective in which "all of this is Maya – illusion."

But don't fight this illusion. Enjoy it for what it's worth. Fighting it would be a mistake.

Let the senses and the mind satisfy themselves by their acceptance of this illusion for the satisfaction of its face-value since they will continue to disagree with the mathematics involved by continuing to emphasize how real is this world of change despite our knowing better.

Philosophy (mathematics) cannot supersede experience. But it can flavor it; or, poison experience if we let it – which will result in an unnecessary state of depression.

It is better to enjoy this illusion than allow ourselves to become depressed about it.

You'll notice that all of the equations of §2xx involve either the square of voltage times negative one, or else the square root of negative one (the imaginary number, "i") times voltage (which is not squared). This signifies the additive inversion (negation) of the phase of voltage by one-half cycle of alternations relative to the phase of current making these equations exclusively relevant to the generation of reactive power and enumerated by complex numbers. This is also signified by the versor algebra[6] operator of "i" for one-quarter cycle of alternations versus the square of "i", or the versor operator of "–1", for one-half cycle of alternations.

A quarter cycle, "i", is due to either capacitive reactance or inductive reactance and is either "+i" or "–i" depending upon the circular direction of displacement within one cycle of alternations is occurring as the result of capacitive reactance shifting voltage backwards by one-quarter cycle, represented by "–i", or else occurring as the result of inductive reactance shifting voltage forwards by one-quarter cycle, represented by "+i".

The generation of reactive power, represented by negative unity power factor, is a shift of voltage by one-half cycle of alternations and is, thus, represented by the square of "i", namely: "–1".

Fortunately, despite the shortcomings of collective confusion surrounding this topic, we can still get a grand total of energy accountability by adding up all of the subtotals of reactive power generation versus all of the subtotals of real power consumption (which electronic simulators provide us) since the summation of this pair of subtotals will be a real number of either negative or positive outcome indicating whether production of energy predominates or else consumption of energy predominates, respectively.

---

6   Versor Algebra, by Eric P. Dollard → http://versoralgebra.com/

Thus, and most importantly, we can discover whether or not a circuit is symmetrical and whether it is thermodynamic, or else is asymmetrical and negentropically or entropically non-thermodynamic. Looking at the equivalency of the absolute values of both subtotals will *suggest* symmetry versus non-symmetry, and is conclusive, because real power is the compliment to conductivity (reactive power). To attempt to add these two parameters together to come up with a grand total of a singular parameter is, also, possible, because each is the multiplicative inverse (integer versus fraction) and additive inverse (positive versus negative) of the other.

The following diagram, on page 14, is of a simple flashlight circuit engaging a mere two components: a battery and a resistive load representing an incandescent bulb. It's a simple example of how a segregated analysis tabulates the power of each component and gives each a sign polarity of positivity, if the outcome is wattage consuming real power, or a polarity of negativity if the outcome is volts/amperes producing reactive power.

This tabulation will give a pretty good idea whether this circuit is symmetrical in which power produced equals power consumed (exported; converted) into a useful byproduct, such as: light, heat, or mechanical motion (in a motor). Or, in the alternative, whether this circuit is asymmetrical in which power produced does not equal power consumed. This latter condition occurs if the power is reactive. Equality of production versus consumption occurs if the power is real.

Ohm's Law is obvious. Mho's Law is not. Mho's Law must be inferred from Ohm's Law (the not-so-obvious must be inferred from the obvious) or, else, we'd have a skewed vision of electrical reality lopsidedly in favor of a well-ordered society predicated upon the constant need to replenish our failure to renew our own electricity instead of "buying" it from our authority figures either as a material product or as the informational and emotional indoctrination which is required to maintain this sham. These figures of authority are numerous and varied, involving: commerce, education, governmental regulation, military defense of these beliefs, entertainment to assure us that this is all OK and not anything to get upset over, etc.

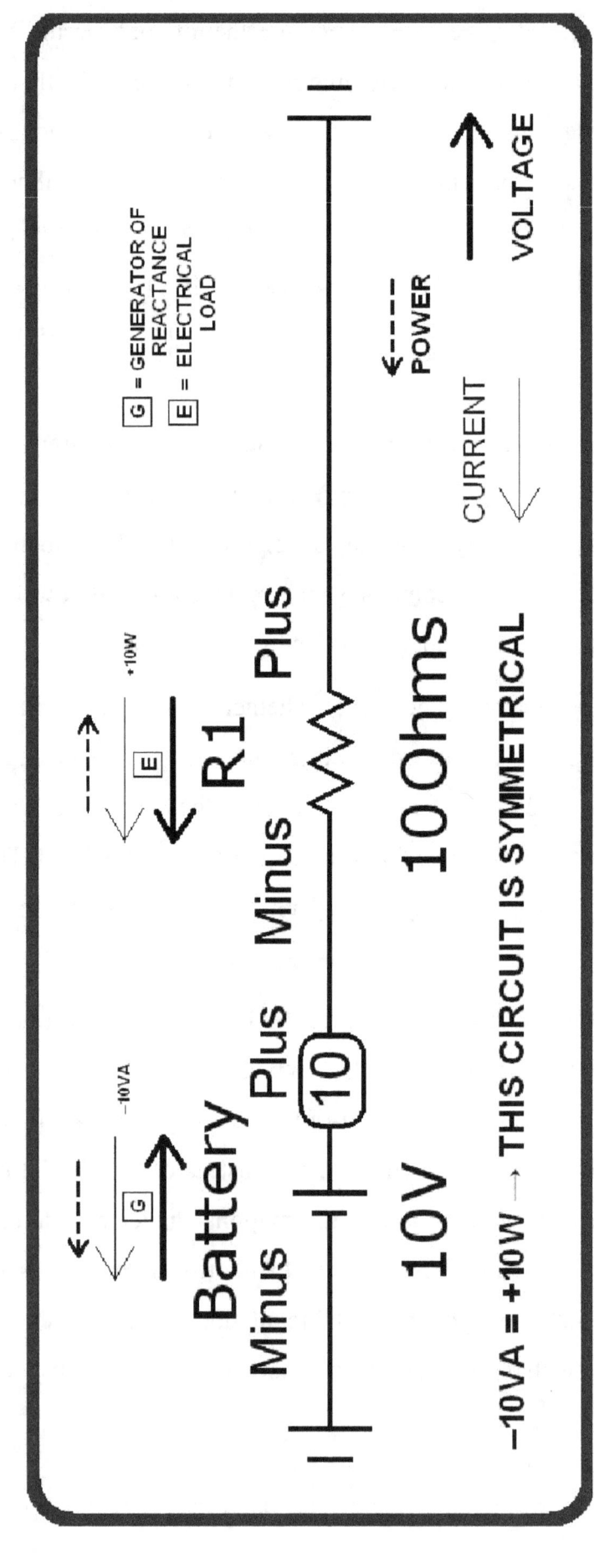

What comes out of a battery is not energy since its current is polarized 180° in opposition to the polarity of its voltage. Instead, what comes out of a battery is information in the form of volts/amperes (VA) also known as: reactive power. The chemistry of the battery is reacting to the closure of a switch causing the chemicals inside of a dry-cell battery to inter-react with each other which they would not have done had the switch (of this type of simple circuit which interconnects the two terminals of a battery) never have been closed.

This chemical reaction is potential power which we measure at the terminals of the battery as being a voltage difference between its two terminals. It becomes reactive power when we connect the two terminals of this battery to a circuit and close a switch to engage the chemicals inside of the battery to react against each other causing current to exit from out of the negative terminal which depletes the voltage difference between its two terminals unless it's a rotary generator in which the generator initiates an increased magnetic and mechanical resistance to whatever is attempting to rotate its shaft.

But the orientation of nomenclature remains intact, namely: the negative terminal of a voltage source is still emitting current of a negative polarity while its positive terminal has a positive polarity of voltage. By definition, this implies reactive power; not watts. This means that energy never exits the battery. Energy only enters the circuit, connected to this battery, if the circuit in question merely consumes power without producing any.

The volts and amperes of every component within a circuit, and the polarity of their sign values, can be accounted for to satisfy a *segregated analysis* of a circuit's activity yielding volts/amperes (VA) or watts indicating the generation of reactive power or the consumption of real power at an electrical load.[7]

This contrivance of our creation of the fiction known as: "current," represented by the capital letter of "I", replaces voltage divided by resistance showing that some of the voltage in the power of a circuit represents a change in voltage due to its resistance within the circuit. The other voltage that we have left alone represents the static voltage which drives this changing

---

7   The Meaning of Unity in Energy Conversion Systems, by James F. Murray, III and Aaron Murakami →
    https://is.gd/zujaqu = https://www.amazon.com/dp/1650183658/

voltage brought about by resistance. This is the reason why we have two voltages in Ohm's Law squared against each other before being divided by resistance.

But this is never the way it is taught to us common folk who don't know any better. Instead, we're taught the derivative of Ohms Law known as: power (in watts) equals voltage times current. This is partly true, but is partly a fiction and an oversimplification of the situation.

If we had known the full situation, we might have stumbled upon Mho's Law on our own. But everything that has been done to dissuade us from that possibility has been incorporated into our common sense style of thinking to discourage us from stumbling onto the mathematical justification for free energy tucked away and hidden in plain sight within Ohm's Law and instinctively derivable as Mho's Law.

This, of course, violates the Conservation of Energy, because the Conservation of Energy does not allow for any derivatives that could undermine its monopolistic authority or counter it's argument. Yet, Mho's Law complements Ohm's Law because they are multiplicative inverses of each other in one sense, and are additive inverses in another sense.

Jim Murray (in his book) encourages us to perform a segregated analysis of all of the components of an overunity circuit, or of a conventional circuit (which is not expected to be overunity), to discover whether or not they are overunity, or underunity, and where (located within each electronic component) is each of these coefficients of performance located? Because some components of a circuit can sometimes produce reactive power without a whole lot of assistance from any external prime mover. And it would be good to find out which components are performing this feat of Herculean strength.

In our pursuit of this endeavor, segregated analyses can also shock us with the appearance that some electronic components may consume power at a rate which is faster than it is being produced elsewhere within a circuit, or outside of a circuit from its power supply feeding that circuit. Surprises are in store for us...!

It turns out that capacitance can become saturated with dielectric charge (measured in volts) to such an extent that capacitance will no longer allow itself to absorb any more electric charge (in the form of voltage). If this state of affairs can be continuously maintained over the course of the alternating cycles of voltage within an oscillating circuit, then a funny thing happens is the removal for the need for any external prime mover serving as the exclusive source of power for that circuit.[8] Instead, coils – all coils including coils intended to serve as electrical loads, such as: the coils of an electric motor – become generators of reactive power despite the fact that none of these coils need to rotate (ie, move) within a magnetic field to satisfy Michael Faraday's Law of Induction[9] which results in the (fictional!) flow of current. *{Keep in mind that current, the flow of current, is a fiction. The reality is the change in the square of voltage, over time and due to resistance, without any current effecting this change. Hence, photons are another fiction not worthy of our serious consideration. Unfortunately, this view is so commonplace that it is impossible to avoid it occurring as a pandemic condition of our sciences.}*

This transcendence of Michael Faraday's Law of Induction is due to the fact that whenever a capacitor maintains saturation, it can no longer absorb voltage and, thus, must reflect this voltage back to its source (what ever that source may be). And this reflection of voltage is immediately simultaneous to its reception requiring no speed of light to transfer this voltage (return it) back to its source since this reflection of voltage is devoid of any current, whatsoever.

Only the current portion of electricity defines the electromagnetic wave of light which is regulated by a definitive speed. Voltage does not travel. It appears, simultaneously, on both sides of a capacitor without any time-lag.

This reversal of voltage effectively cancels the application of voltage which was initially applied to a saturated capacitor resulting in zero watts (and zero current) reflected from this capacitor. This reflection of voltage constitutes a standing wave – in as much as, the phase of

---

8 Please see a self-published copy of my application for a provisional (temporary) patent, at →
   https://is.gd/ovikam = https://www.amazon.com/dp/B098H61T78 ← Vol. 1, text
   https://is.gd/mirudo = https://www.amazon.com/dp/B098GJDG7D ← Vol. 2, figures

9 Faraday's law of induction → https://is.gd/opamoc =
   https://en.wikipedia.org/wiki/Faraday%27s_law_of_induction

applied voltage is one-half cycle of 180° of separation from the phase of the other voltage...the voltage which changes over time due to resistance.

This voltage (which changes over time due to resistance) we, more often than not, call: "current" while the other voltage, which does not change (yet provides pressure or suction applied to the voltage which changes) remains static.

Since this reactive power cannot leave, ie. exit, the circuit, it bounces around inside of it and also accumulates at an escalating rate of hyperbolic increase towards the ultimate goal of annihilation of the circuit if it is not curbed, in some manner or another, from reaching infinite oblivion.

Thus, is Mho's Law stated as: resistance divided by the negation of the square of voltage...

$$Conductivity = \frac{Resistance}{-Voltage^2} \qquad §2.1a$$

And, although this defines conductivity, this conductivity has a direct impact on the buildup of reactive power within a circuit.

Reactive power, although lossless, is not useless. Don't let formally trained (ie, indoctrinated) electrical engineers fool you into believing in their own despondent point of view.

It is easy to envision how to convert reactive power into real power and, thus, make use of it by taking advantage of several methods for its conversion, namely: to pass it through a resistor to heat up water and run a steam engine off of this free production of heat energy, or a coil could be wound with two windings in which each winding is counter-wound with respect to the other winding so that the phase of current of one winding may engage the phase of voltage of the other winding, and vice versa, to get useful watts out of this wattless condition. This is accomplished through the union of magnetic fields mutually shared between these two coils.

By the way, the reversal of the phase of voltage of a saturated capacitor is not intuitively straightforward. It results from the simultaneous deflection of two capacitances erupting out of a saturated capacitor at 90° of opposing directions of deflection.

So, if one deflection erupts out of the capacitor to one side, then its complimentary deflection simultaneously erupts out of the capacitor to its opposite side without any time-lag.

Each of these two deflections are mathematically represented by the imaginary value of the square root of negative one.

Notice, how this imaginary value is represented by the lower case letter of "i." It's as if the capital letter of "I," which represents the fictionalized existence of current, has replaced the more realistic representation of reactive power which is represented by the lower case, letter "i." Here is another instance of how Mho's Law has been swept under the proverbial rug of our collective ignorance.

Their squaring when combined (since they occur, simultaneously) results in negative one placed alongside (multiplied against) the square of voltage in the denominator of Mho's Law underneath (dividing into) resistance located in the numerator of Mho's Law...

$$Conductivity = \frac{Resistance}{-Voltage^2} = \frac{1}{Voltage\sqrt{-1}} \times \frac{Resistance}{Voltage\sqrt{-1}} = \frac{Current}{Voltage\sqrt{-1}} \qquad §2.1ab$$

In other words, whenever a saturated capacitor reflects the square of voltage, it does two things at the same time: it tangentially reflects one imaginary value for voltage, but it also allows for the tangential reflection of the other value of imaginary voltage in opposition to the first tangential reflection of voltage. And their simultaneity combines to result in a negation of voltage outside of the field of complex numbers, yet, within the realm of real numbers.

This negation of the square of voltage is vital for identifying the condition for which free energy occurs, namely: the eradication of any preponderant dependency upon external prime movers to power our circuits.

As an aside, this ideology has another consequence...

Besides the severe reduction of a circuit's dependency upon any external prime mover to power itself, Mho's Law encourages (by way of its stipulation) that having a high resistance actually *encourages* the production of free energy whenever voltage gets inverted by one-half cycle of its phase relation with current. This is due to the variable of resistance is located in the *numerator* of Mho's Law. Furthermore, the location of the square of voltage in the denominator of Mho's Law encourages us to use very little voltage to act as the circuit's traditional prime mover held outside of our circuit if we are to achieve maximum conductivity at room temperature.

Although I'd like to avoid the use of the word: "current" in this discussion, I can't help to entirely avoid it since all of our measuring devices and electronic simulators make frequent use of the term of: "current" as if it were a solid fact of electrical engineering. So, be it...

It is worthwhile to mention that there are innumerable methods for saturating a capacitor to yield the reversal of voltage. But they all share one thing in common, that they occur within the context of an elevated inductance.

Let me explain...

This negation of voltage is a trademark of a common feature of coils, namely: their back EMF. It turns out, that this reversal of voltage cannot occur all by itself within the context of a saturated capacitor to be of any use, namely: be able to power a load. Instead, this saturated capacitor must be supported by a very large inductance yielding a very large back EMF. It is this enlarged back EMF that will encourage a saturated capacitor to yield a reversal of voltage and not collapse the moment we apply an external load. This enlargement of back EMF can be produced in a number of ways...

We could use a very large coil, such as a coil of thousands of Henrys of induction, or...

We could use a permanent magnet of very strong magnetism, or...

We could enlarge the mass of ferromagnetic material associated with an inductor while not necessarily increasing its inductance or its magnetism.

This latter condition was used by Tesla, and cited by a Mr. Dort who was the son of his father who had worked with the Germans in implementing Tesla's Special Generator to recharge their batteries in some of their Electro-U-Boats of WWII – according to William Lyne. Lyne quotes Tesla as claiming that, "for every two hundred pounds of iron added to his Special

Generator, its output was increased by one horsepower."[10]

We could use powerful permanent magnets available to us, today, but not available over 100 years ago (in 1893) when Tesla displayed his Special Generator for the first time to a small audience as documented by Thomas Commerford Martin in his book on Tesla's inventions.[11]

Or, we could use a very large coil of iron wire surrounding a large iron bolt in Nathan Stubblefield's Electric Battery patent.[12]

But Tesla had to use the most efficient method that was available to himself at that time. He chose to enlarge the ferromagnetic mass which was magnetically coupled to the core of his Special Generator to make up for its lack of induction at no additional cost of resistance.

How can this be?

Magnetic remanence and the coercivity of permanent magnets helps to explain how Lyne's quotation of Tesla is a valid, scientific fact.

In other words,...

Paul Babcock displayed a huge Perpetual Motion Holder[13] (operating under the principles of magnetic remanence) at the Science, Energy and Technology Conference in Hayden, Idaho, back in 2013. Its core was so large, that it became very difficult to pull apart this core into its two parts of a steel bar resting at the feet of a horseshoe-shaped core without resorting to the use of a sheering force sliding the steel bar sideways to affect a separation. Even then, the use of a considerable sheering force (at right angles to the plane of the horseshoe) was needed and remained difficult to separate the core material into its two parts while the core was allowed to

---

10 PENTAGON ALIENS; CHAPTER VIII: A TASTE OF OTHER ENERGY SECRETS → https://is.gd/anavum = https://www.bibliotecapleyades.net/ciencia/pentagonaliens/pentagonaliens08.htm#CHAPTER%20VIII:%20A%20TASTE%20OF%20OTHER%20ENERGY%20SECRETS

11 https://is.gd/spec_gen = http://vinyasi.info/circuitjs1/texts/Nikola Tesla/The Inventions, Researches and Writings of Nikola Tesla, ch. 63.pdf

12 Nathan B. Stubblefield, Electrical Battery; patented 8 March 1898; #600,457 → https://is.gd/atiyim = http://vinyasi.info/circuitjs1/texts/Nathan%20Stubblefield/US600457.pdf

13 Look it up on YouTube. Edward Leedskalnin built Coral Castle in Florida and popularized the Perpetual Motion Holder although he did not invent it.

retain its memory of having become charged up. Once it released its memory as a short burst of energy, it totally forgot to adhere the two pieces of core together.

Paul powered his demonstration circuit with a momentary spark arising from nothing other than a mere 9V battery!

It was the largess of his massive core which made all the difference. And he knew this, also. He did not make his demonstration model so large by happy, coincidental accident. He knew what he was doing!

He was overcoming the resistance of inductance by compensating it with a massive ferromagnetic core which possesses no resistance, nor any impedance, to its magnetic remanence at any size.

Yet, the more massive is this core, the greater is the momentum of whatever magnetic memory is stored there. And this increased momentum contributes to overcoming the inertia of inductance.

Thus, the inertial inductance is fixed by the size of the coil which is wound upon a ferromagnetic core. But the magnetic remanence is not fixed. It varies based on the mass of ferromagnetic material which is magnetically coupled to the core of the inductor.

The sky is the limit!

◦ ◦

The following page, 23, is the original photo of the two Ammann brothers standing in front of their 1921 electric vehicle conversion in Denver, Colorado. It was already an electric car. They converted it to run without any batteries and substituted their own power supply strapped to the front end of their vehicle. Page 24 contains this original photo in a newspaper article.

Someone has superimposed two arrows, upon the photo of page 23, which points towards the two mysterious spheres which the brothers placed into the car's headlight sockets after removing the headlights.

The following images are screenshots of my second attempt at replicating the Ammann brothers' mysterious invention by using a spark gap to induce the inversion of voltage whose phase is totally opposite to the phase of current by one-half cycle of alternating voltage possessing no prime mover other than a capacitor, C1, precharged with one volt. This induces the mathematical definition of the generation of reactive power (invoking Lord Kelvin's suggestion of using Mho's Law) at all three inductors, L1 & L2 & L3, since their assumed wattage is actually negative volts/amperes (not positive watts) on page 28. Micro-Cap[14] simulator doesn't know any better to differentiate between generators of reactive power outputting negatively signed polarization (of power) versus consumers of real power outputting positively signed polarization. So, we'll have to reinterpret the results now that we know better!

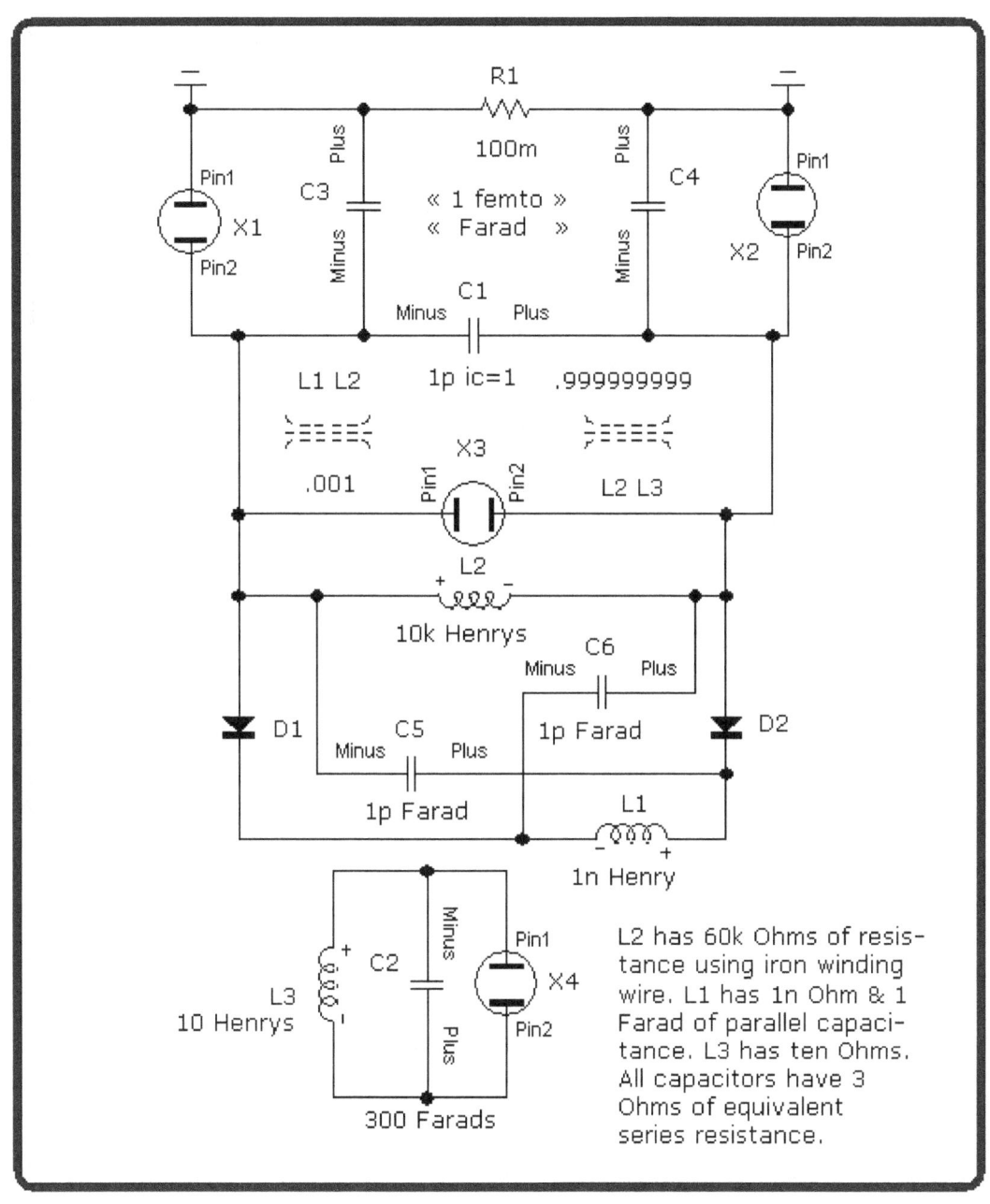

---

14 http://spectrum-soft.com/

Here are the nodal voltages of the circuit schematic, up above...

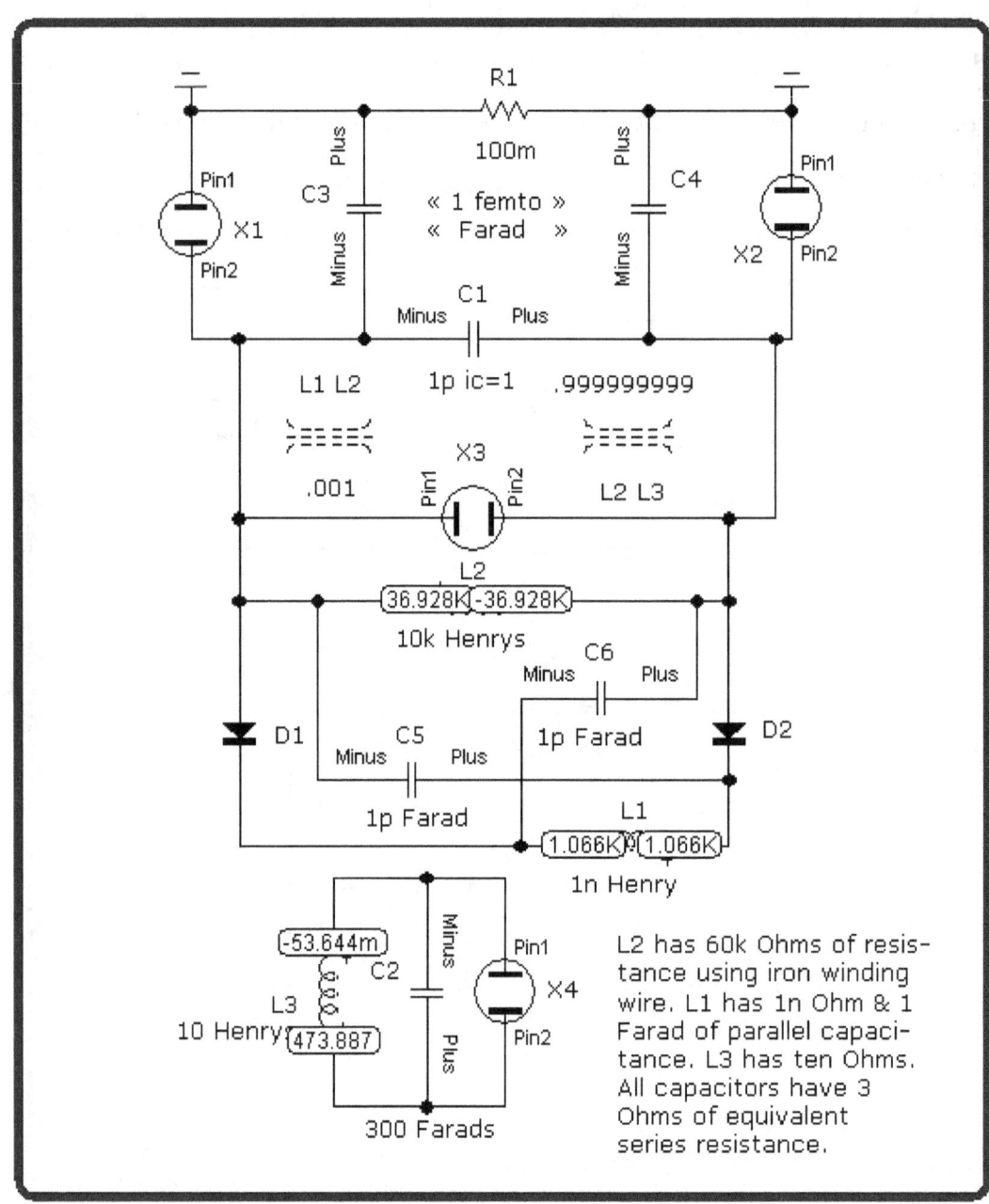

The Ammann brothers mystery shall remain just that...a mystery, along with the mystery of Nikola Tesla's TriMetal Generator. But we'll continue to plunge ahead with our own attempts at doing something similar...

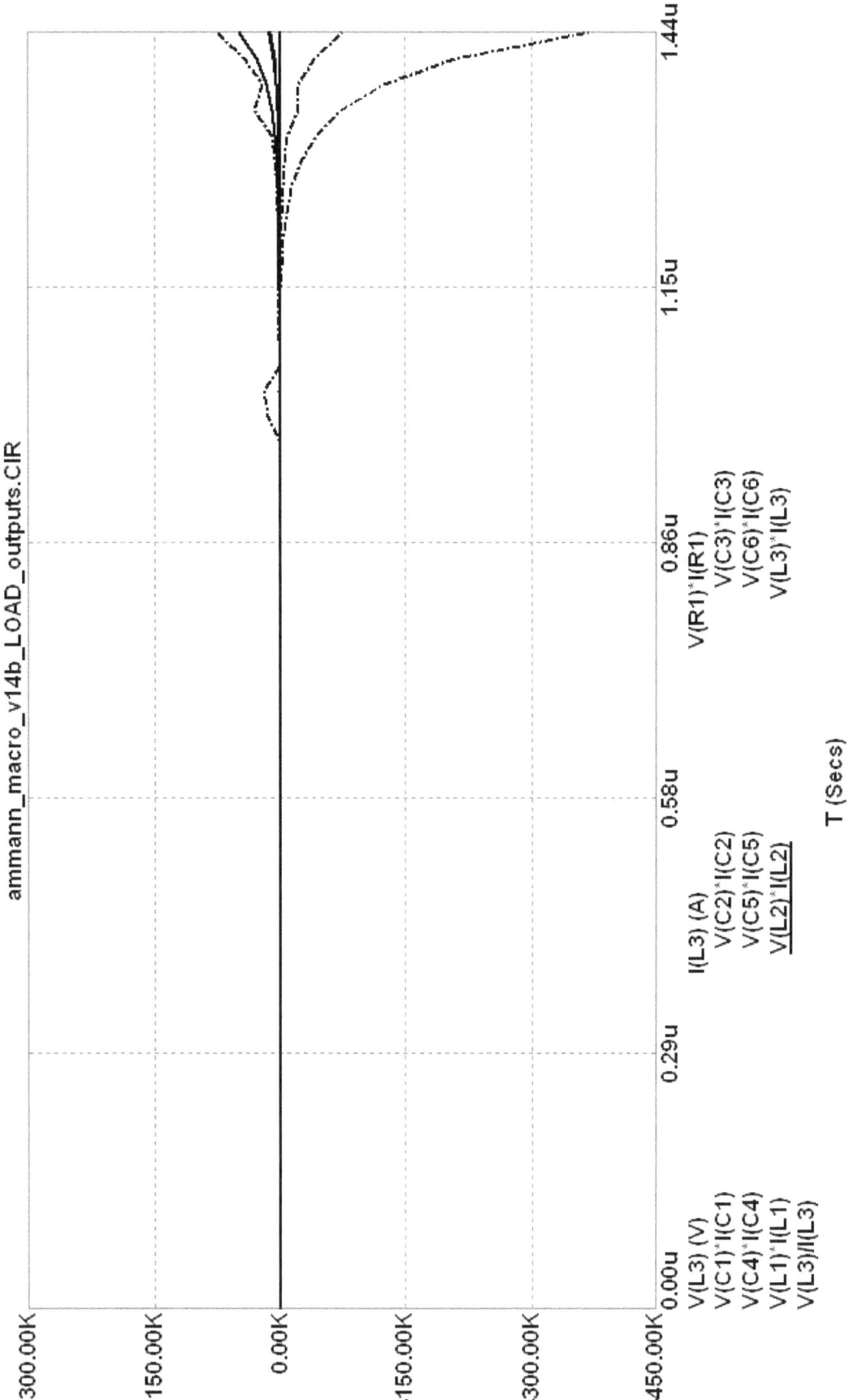

ammann_macro_v14b_LOAD_outputs.CIR

HERE IS A EXCELLENT EXAMPLE OF POSITIVELY SIGNED WATTAGE (AKA, REAL POWER) VERSUS NEGATIVELY SIGNED VOLTS/AMPERES (AKA, REACTIVE POWER). THE NEGATIVELY SIGNED REACTIVE POWER IS ARISING FROM ALL THREE INDUCTORS. L1 AND L2 AND L3 HAS REACHED ITS TARGET OF AROUND 70,000 VOLTS/AMPERES TO EMPOWER AN ELECTRIC MOTOR DRIVING AN ELECTRIC CAR IF ITS WINDINGS ARE BIFILAR AND CROSS-WOUND SO THAT THE PHASE OF VOLTAGE OF ONE WINDING WILL MATCH (AND COMBINE WITHIN THE DOMAIN OF THEIR MUTUAL INDUCTANCE) THE PHASE OF CURRENT OF ITS COMPLIMENTARY WINDING AND VICE VERSA TO (HOPEFULLY!) INDUCE AN EMISSION OF WATTAGE.

A NEGATIVE VOLTAGE TIMES A POSITIVE AMPERAGE YIELDS A NEGATIVE VOLTS/AMPERES, AKA:

	Left	Right	Delta	Slope
V(L3) (V)	242.113	−473.941	−716.053	−1.766G
I(L3) (A)	4.989	157.985	152.995	377.358M
V(R1)*I(R1)	0.000	0.000	0.000	0.000
V(C1)*I(C1)	332.501m	49.084K	49.084K	121.064G
V(C2)*I(C2)	19.540K	74.873K	55.334K	136.479G
V(C3)*I(C3)	83.268u	12.271	12.271	30.267M
V(C4)*I(C4)	83.113u	12.272	12.271	30.267M
V(C5)*I(C5)	−4.302	11.917K	11.921K	29.403G
V(C6)*I(C6)	4.469	12.625K	12.621K	31.129G
V(L1)*I(L1)	−237.243n	−604.321u	−604.083u	−1.490K
V(L2)*I(L2)	−6.366	−368.983K	−368.976K	−910.068G
V(L3)*I(L3)	−1.208K	−74.875K	−76.083K	−187.657G
V(L3)/I(L3)	48.529	−3.000	−51.529	−127.093M
T (Secs)	1.034u	1.439u	405.438n	1.000

REAL POWER — WATTS
REACTIVE POWER — V.A.

Here is a schematic of the same circuit turned OFF by shorting out one of its coils, L2, to ground...

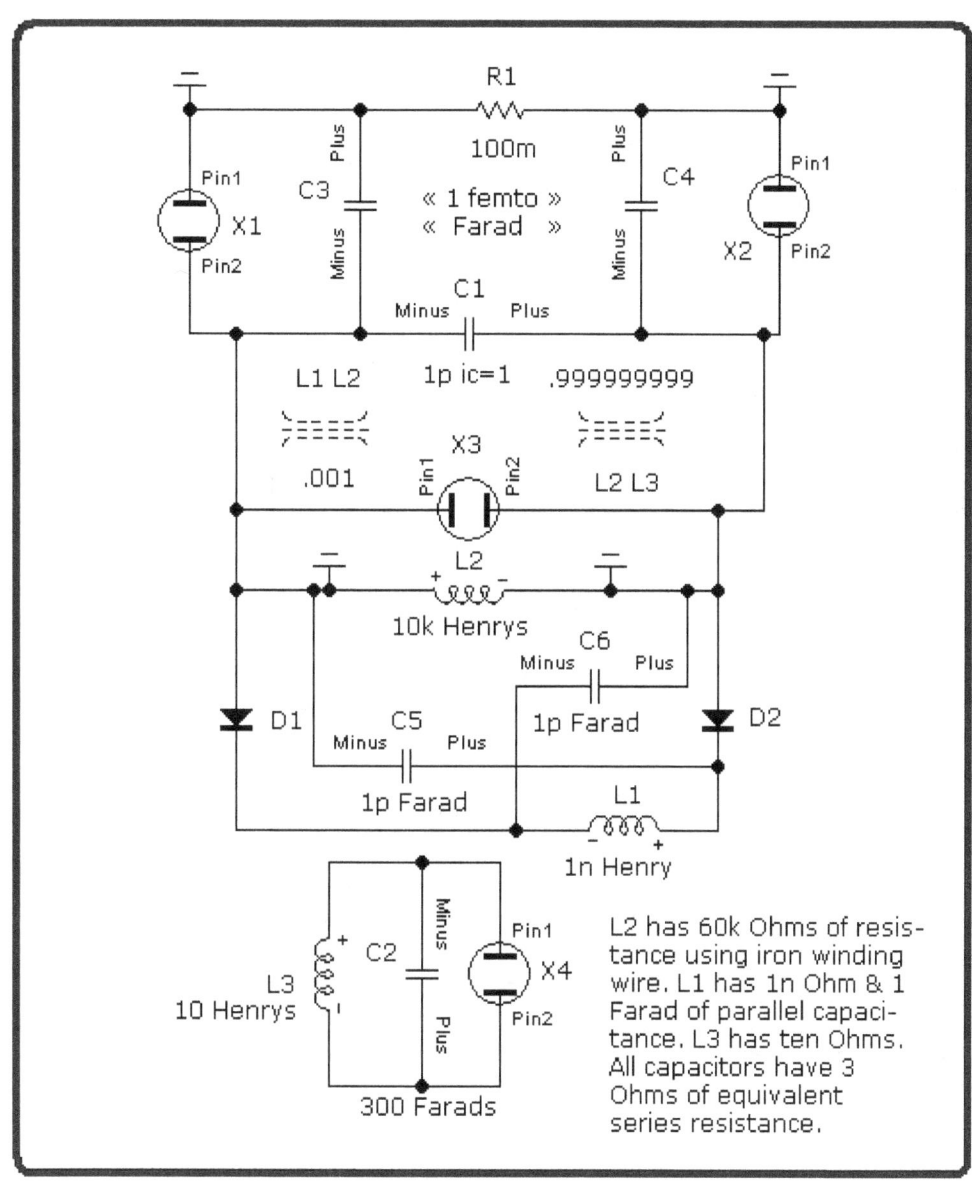

Here are its nodal voltages...

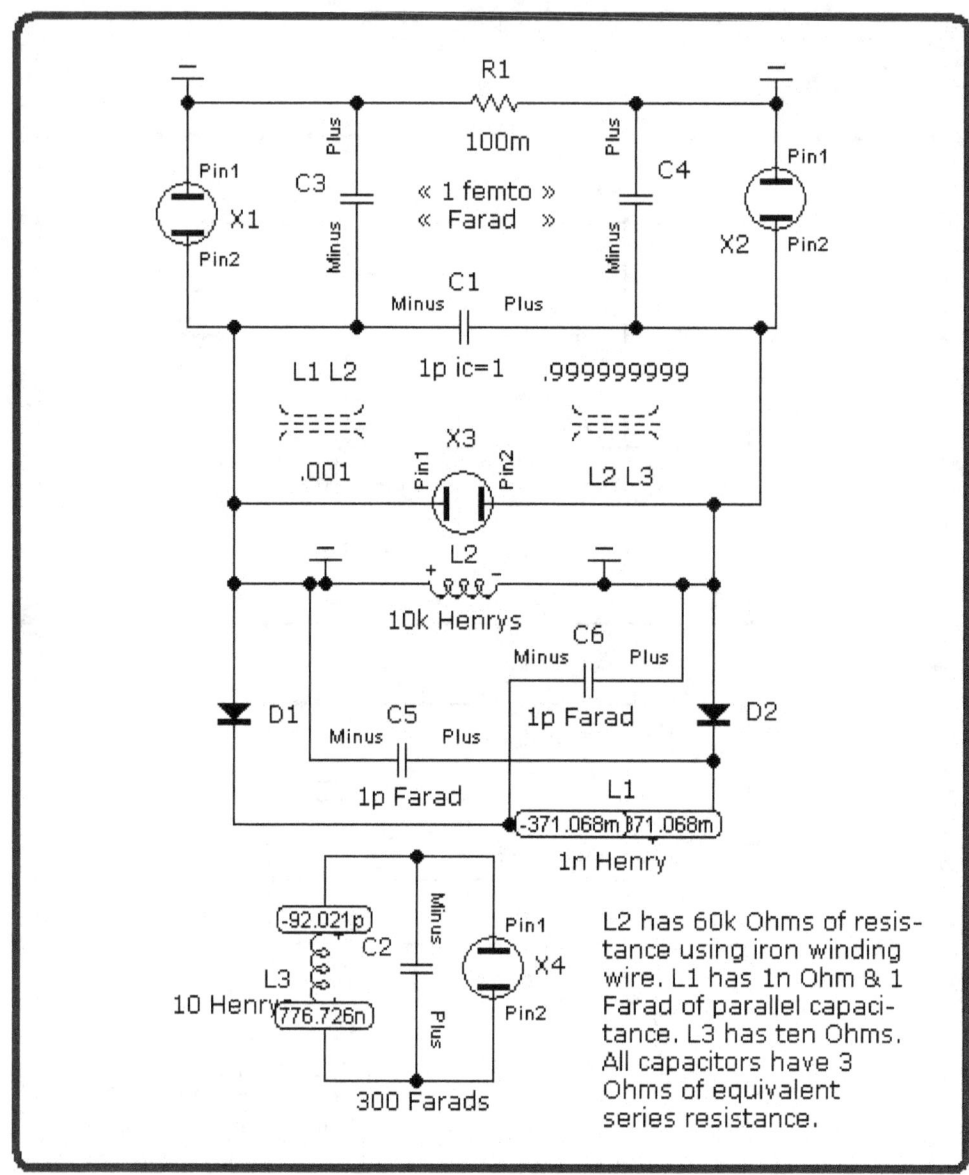

The following page is a screenshot of some of the outputs of various components hovering around zero showing how (relatively) OFF is this shorted circuit...

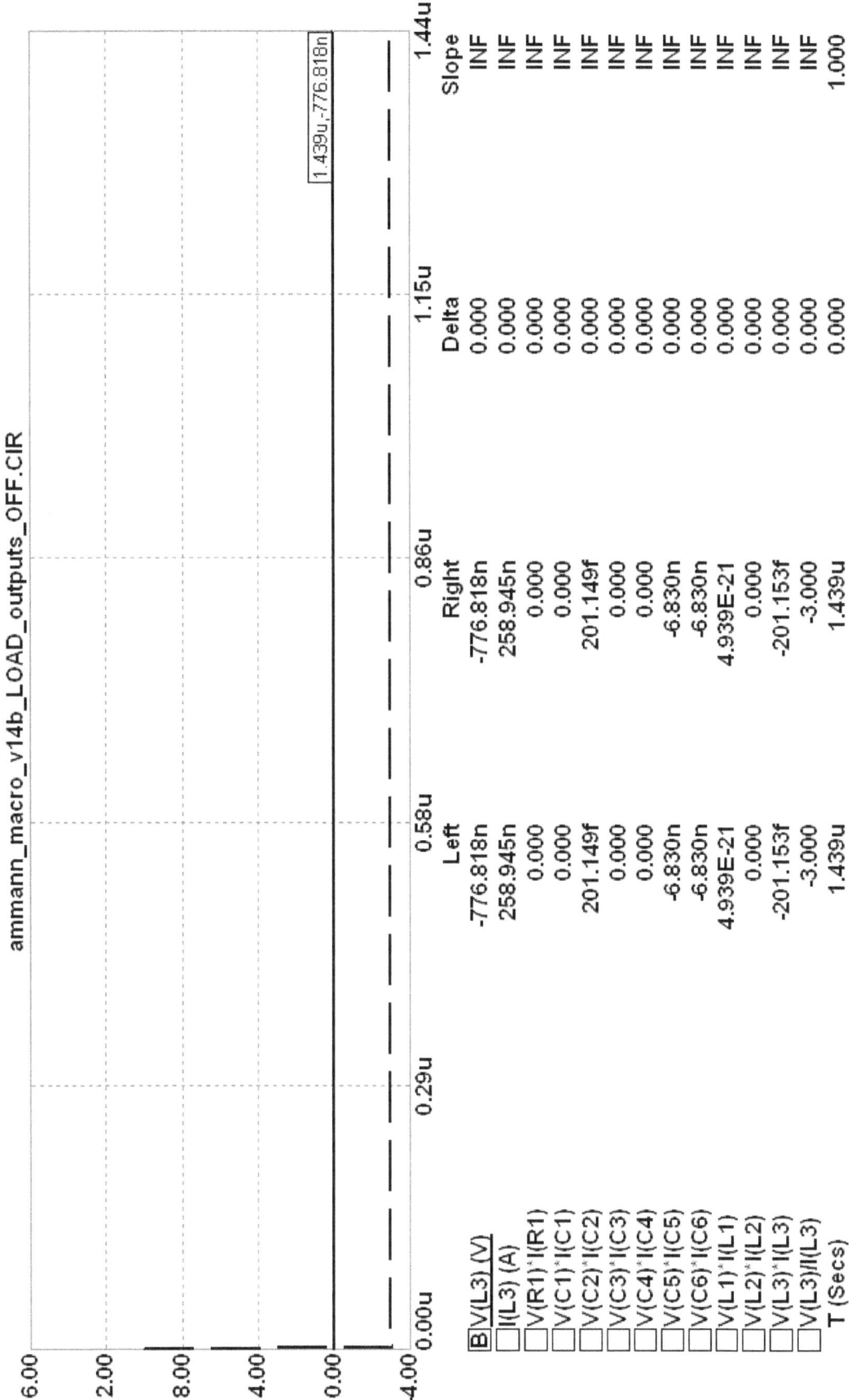

www.ingramcontent.com/pod-product-compliance
Lightning Source LLC
Chambersburg PA
CBHW080818220526

45466CB00011BB/3612